NEW TOPICS FOR SECONDARY SCHOOL MATHEMATICS
Materials and Software

GEOMETRIC PROBABILITY

Department of Mathematics and Computer Science
North Carolina School of Science and Mathematics

NATIONAL COUNCIL OF TEACHERS OF MATHEMATICS

Library of Congress Cataloging in Publication Data:

New topics for secondary school mathematics.

Contents: [1] Geometric probability.
1. Mathematics--1961- . I. North Carolina School of Science and Mathematics. Dept. of Mathematics and Computer Science.
QA37.2.N49 1988 510 88-5305
ISBN 0-87353-258-9 (set)
ISBN 0-87353-259-7

Printed in the United States of America

Contents

Preface

These materials on geometric probability compose the first unit in a course being developed by the Mathematics Department at the North Carolina School of Science and Mathematics. This course, which is being developed under funding from the Carnegie Corporation of New York, is called Introduction to College Mathematics and is designed to prepare high school students who have completed Algebra 2 for the variety of math courses they will encounter in college.

The content and focus of the course respond to recommendations about secondary mathematics that have been made by the College Board, the National Council of Teachers of Mathematics, the Sloan Foundation, and the National Science Foundation. There is an emphasis on problem solving that involves combinations of skills, and real-world applications are used to stimulate the study of some traditional precalculus material. The course goes beyond the traditional precalculus curriculum to provide students with mathematical experiences that will prepare them for future courses and careers. These experiences include analysis of data, an introduction to algorithms, work with discrete phenomena, and exposure to mathematical modeling. The course also makes extensive use of calculators and computers; its design is based on the assumption that a single microcomputer is available for use in front of the classroom.

The *Geometric Probability* unit sets a tone for the rest of the course. By beginning the school year with new, different, and thought-provoking problems, we alert students to the fact that this course is not a traditional precalculus course. They will see familiar mathematics in new contexts and learn how to apply math in sometimes surprising ways. They will also be exposed to some creative aspects of doing mathematics.

Although the unit was written to serve as an introduction to the ICM course, it is well suited to stand alone. Assuming only a knowledge of linear functions and of areas of simple geometric regions, the unit shows students how to determine probabilities of events for which the number of possible outcomes is infinite.

Computer software and a user's guide have been written to accompany the unit. The software provides simulations of seven of the experiments discussed in the unit. These simulations help students understand the random nature of the experiments and appreciate the relationship between empirical and theoretical probabilities.

The unit is somewhat independent from the rest of the course in terms of content development but not in terms of problem-solving approaches. Students learn that there are few hard and fast rules governing problem solving and that a single problem can be treated at different levels of sophistication. The unit also provides an introduction to mathematical modeling, computer simulations, and the distinction between discrete and continuous phenomena.

Many of the ideas and problems discussed in this unit have previously been presented by Richard Dahlke and Robert Fakler in articles published by NCTM. We have used these ideas as a point of departure for further development of problem-solving approaches and mathematical modeling.

Geometric Probability

1 Introduction

The underlying principles of geometric probability are quite simple and straightforward. Consider the targets represented in Figure 1.

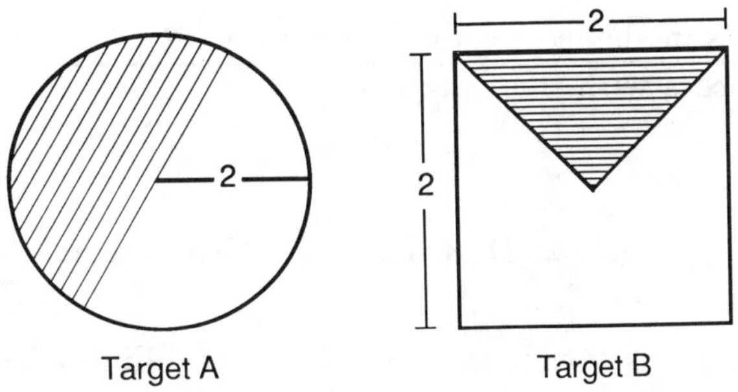

Figure 1: Two Targets

Suppose a dart is thrown at each target so that it is certain to hit the target but the thrower cannot control where on the target the dart lands. That is, the dart is thrown randomly and it is equally likely that it will land on any region of the target. What is the probability that the dart lands in the shaded region? The region that represents all possible points where the dart can land is often called the *sample space*. In this example, the sample space is the entire surface of the target. For target A, the sample space has area 4π, and for target B the area is 4. The region consisting of points that represent a successful outcome is called the *event space*. In our two targets, these are the shaded regions. The *probability of a success* (that is, of landing in a shaded region) *is simply the ratio of two areas*, the area of the event space and the area of the sample space. Since in target A the shaded region is one-half of the total area of the target, the probability of landing in the shaded region is $1/2$. We can see that this is the ratio of the two areas: $2\pi/4\pi$. In target B, the shaded region represents only one-fourth of the total area so the probability is $1/4$.

This unit was developed by the Department of Mathematics and Computer Science at the North Carolina School of Science and Mathematics through a grant from the Carnegie Corporation of New York.

What does it mean for the probability to be 1/4? One-fourth is the ***theoretical probability.*** If we were actually to perform the experiment (that is, actually throw darts), what would we expect to happen? If you throw 4 darts, will one and only one dart hit the shaded region? If you throw 100 darts, how many do you expect to land in the shaded region? Computer software has been provided that simulates randomly throwing darts at a target like the one on the right in Figure 1. When a simulation is run, the fraction of the darts that actually land in the shaded region is sometimes called an ***empirical probability.*** Clearly, you expect the empirical and theoretical probabilities to be close in size, but they will not necessarily be equal.

Class Practice

1. Use the software to simulate throwing 100 darts. Repeat this several times. What was the largest number of darts that landed in the region? What was the smallest? On average, how many of the darts landed in the shaded region?

2. Use the software to simulate throwing 10 darts. Repeat this experiment 20 times. Complete Table 2. Discuss any observations you can make about the number of darts that land in the shaded region.

Number of darts in shaded region	0	1	2	3	4	5	6	7	8	9	10
Number of experiments with this outcome											

Table 2: Dart Throwing Experiment

3. Explain what it means to say that the probability of rolling a sum of seven on a pair of dice is 6/36.

4. Suppose you are playing a game of chance with a friend and the probability of your winning is 0.4. Tell what you expect to happen in each of these situations:

   (a) You play the game once.

   (b) You play the game 100 times.

   (c) Suppose it costs you $1 every time you lose a game. How much money would you need to gain each time you win in order for it to be worth your while to play?

Given the target in Figure 3, what is the probability that a randomly thrown dart lands inside the square but outside the circle? The sample space is the square target, which has area 25. The event space is the region outside the circle, which has area $25 - 25\pi/4 = 25(1-\pi/4)$. The probability that the dart lands inside the square but outside the circle is

$$\frac{25(1-\pi/4)}{25} = 1 - \pi/4 \approx 0.22.$$

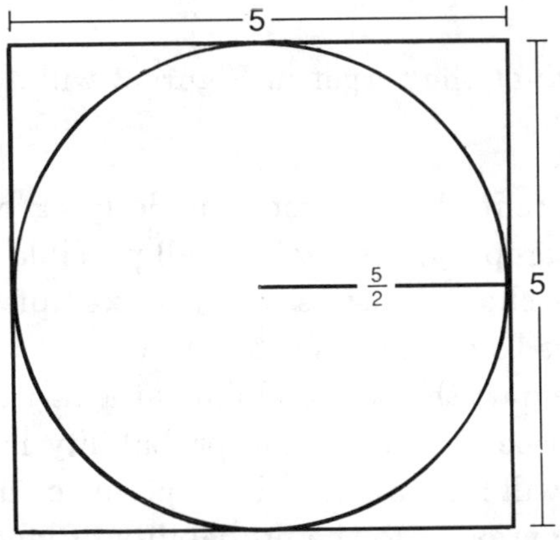

Figure 3: Target

2 The Parachute Problem

A parachutist jumps from an airplane and lands in a square field that is 2 kilometers on each side. In each corner of the field there is a large tree (see Figure 4). The parachutist's ropes will get tangled in the tree if she lands within 1/11 kilometer of its trunk. What is the probability that she will land in the field without getting caught in a tree?

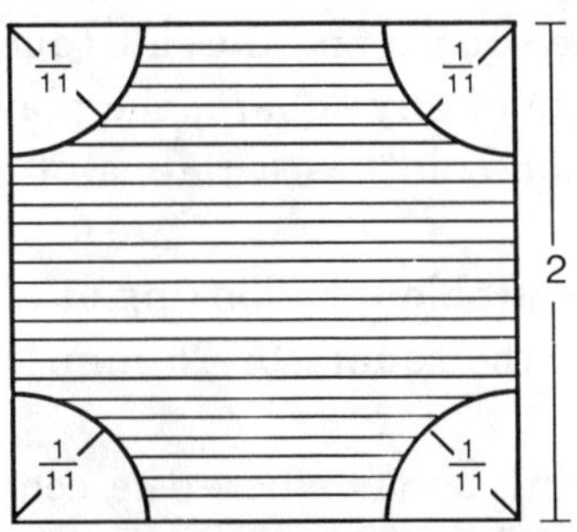

Figure 4: Parachute Problem

In order to find the probability, we must make two assumptions. First, we assume that the parachutist lands somewhere within the field; she does not land outside it. Second, we assume that her landing point is random.

Under these assumptions, the probability of not getting caught in a tree is a ratio of areas. The sample space is the square field — its area is 4 sq km. The event space is the shaded area — its area is $4 - (\pi/121)$. The probability of landing in the shaded area is $(4 - (\pi/121))/4$, or approximately 0.99.

Class Practice

1. Rework the parachute problem, using a radius of 1/9 instead of 1/11. Now try a radius of 1/5.

2. Explain how throwing darts at the target in Figure 3 will allow you to estimate the value of π.

You should notice two important characteristics underlying both the target problem and the parachutist problem. First, the sample space contains all possible outcomes. Second, the outcomes occur randomly. Watch for these characteristics in other examples — they are essential if you are going to use geometric techniques to compute probabilities.

What is the probability that a dart thrown randomly at a dart board lands on any one particular point? The area of the event space is zero, so the probability is zero. For the same reason, the probability that the parachutist will land on any single point in the field is also zero. Likewise, the boundaries of the trees have zero area — so the probability of landing on these boundaries is zero. Does it make sense to you that the probability is zero? Does this mean that it is impossible to land on any single point? There is potential for some confusion here — it arises from our difficulty in grasping just how big infinity is and just how small a point is. If there were a trillion points in the target, then the probability of landing on any particular point would be one-trillionth. However, there are not a trillion points on the square; there are an infinite number. The probability of landing on any single point is actually zero.

3 The Fairground Problem

At a state fair a game is played by tossing a coin onto a large table ruled into congruent squares. If the coin lands entirely within some square and doesn't touch any edges, the player wins a prize. If the coin touches or crosses the edge of any square, the player loses (see Figure 5). Suppose the squares are S units on a side and the coin has a radius of R units. What is the probability of winning this game?

First let's get acquainted with the problem by looking at a few specific examples. Let the radius of the coin be 10 mm and the side of the square be 25 mm. What geometric shapes represent the sample and event spaces?

The key to this problem is to think about where the center of the coin lands. If the distance from the center of the coin to the edge is less than or equal to 10 mm, the coin will touch the edge of a square. So the center of the coin must be more than 10 mm from each side in order for the player to win a prize (see Figure 5).

Throwing a coin at random corresponds to randomly picking a point within the square, with the point representing the center of the coin. A successful throw corresponds to choosing that random point inside a 5×5 square in the center of the larger square. Speaking in terms of geometric probability, the sample space is the square with side 25 mm and the event space is the square with side 5 mm. The ratio of these two areas is $25/625 = 1/25$. The probability of winning this particular game is $1/25$.

If we change the side of the square to 20 mm, what is the probability of winning? Now the sample space has an area of 400. The event space has an area of 0. Why? It is impossible to win at this game. Do you understand why?

Now that we have looked at a couple of specific examples, we are ready to consider the general problem: a square of side S and a coin of radius R. To win, the center of the coin must be at least R units from the side. The area of the sample space, the big square, is S^2, and the area of the

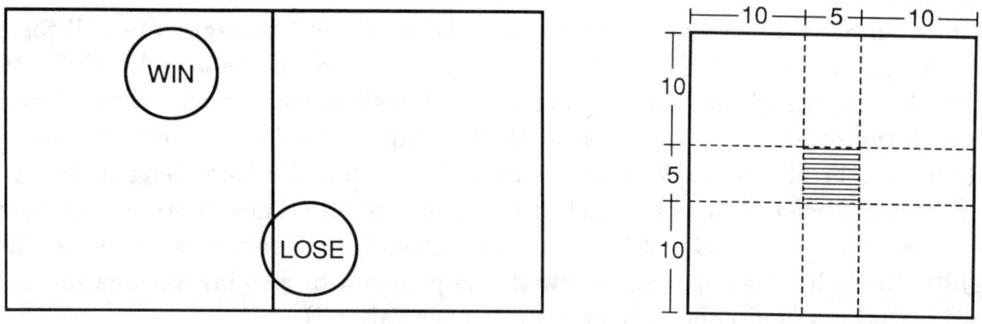

Figure 5: Fairground Problem

event space, the small square, is $(S-2R)^2$ (see Figure 6). The probability of winning such a game is given by

$$P = \frac{(S-2R)^2}{S^2}$$

where S is the length of the side and R is the radius of the coin. Notice that the conditions of the problem imply that R must be less than or equal to $S/2$ so that $S - 2R$ will be nonnegative.

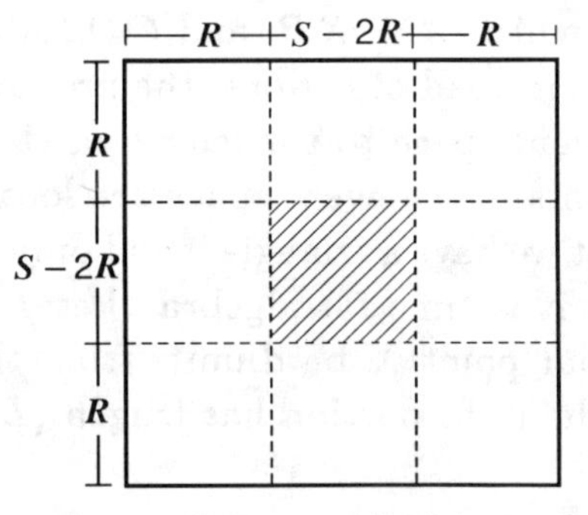

Figure 6: Event Space for Fairground Problem

Class Practice

1. Suppose the square is 50 mm on a side. What should be the radius of the coin so that the player has a 50% chance of winning?

2. Under what conditions on S and R is the player guaranteed to lose the game? Under what conditions would he be guaranteed to win?

4 The Triangle Problem

The students in an geometry class are considering when three line segments will form a triangle. They know that a triangle will be formed if and only if the sum of the lengths of any two segments is greater than the length of the third segment. The class has two wooden dowels, one of which is twice as long as the other. The students break the longer dowel at a random spot, so that they now have three lengths of dowel. What is the probability that the three lengths form a triangle?

The students construct a simple model of this experiment. They represent the dowels by line segments AB and CD, with $AB = 2(CD)$. In this model, the lengths were chosen as $AB = 200$ and $CD = 100$. Next they become familiar with the problem by picking various spots X along AB as their breaking points, and constructing a table (see Table 7).

Choose X so that $AX =$	Length of $XB =$	Length of $CD =$	Form a Triangle?
10	190	100	NO
80	120	100	YES
150	50	100	NO

Table 7: Experimental Results for Triangle Problem

The students realized that some points X along the segment AB work whereas others do not. The sample space of this problem is the set of all points on the line segment AB. The event space is the set of points X on AB for which AX, XB, and CD form a triangle. Although the sample and event spaces are line segments instead of regions, the probability is still a ratio. In this case, it is the ratio of the length of the event space to the length of the sample space. The students have not yet determined where the points in the event space are located.

From geometry, they know that we have a triangle if and only if $AX+XB > CD$ and $AX+CD > XB$ and $XB + CD > AX$. Now it is time for algebra. Let L represent the length of AB. Then $CD = L/2$. Let the position of our point X be d units from the left endpoint of AB, so the left portion of AB has length d and the right portion has length $(L - d)$.(See Figure 8.)

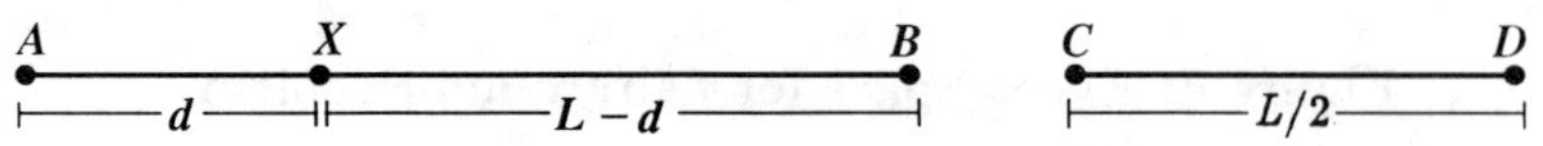

Figure 8: Geometric Model for Triangle Problem

We can now restate our inequalities, substituting $AX = d$, $XB = (L - d)$, and $CD = L/2$.

$$AX + XB > CD \quad \text{and} \quad AX + CD > XB \quad \text{and} \quad XB + CD > AX$$
$$d + (L - d) > L/2 \qquad d + L/2 > (L - d) \qquad (L - d) + L/2 > d$$

Simplifying each of these gives the following inequalities:

$$L > L/2 \quad \text{and} \quad d > L/4 \quad \text{and} \quad d < 3L/4$$

What do these statements tell us? The first is trivially true and gives us no information. However, the second and third inequalities tell us that for the point X to produce a triangle, it must fall between $L/4$ and $3L/4$. The event space is shaded in Figure 9.

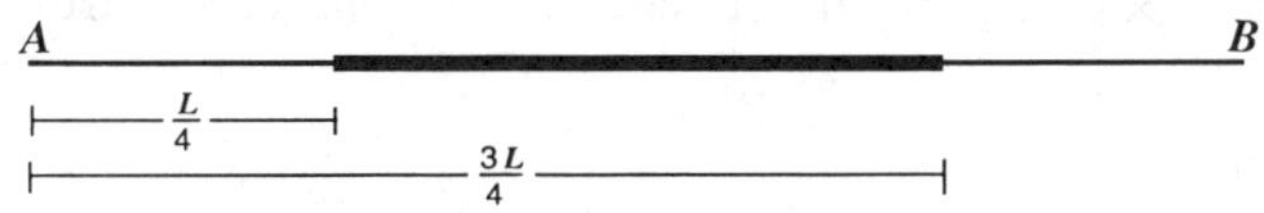

Figure 9: Event Space for Triangle Problem

What portion of the sample space is the event space? Since 1/2 of AB consists of breaking points X that will result in a triangle, the probability must be 1/2. The same answer is obtained by reasoning that the length of the event space is $L/2$ and the length of the sample space is L, so the ratio of these two lengths is 1/2.

Notice the two essential characteristics that allowed us to model this problem geometrically. First, the breaks all occur somewhere on the dowel. Second, the breaks occur at random points. According to our model, breaking a dowel at a random point is equivalent to randomly choosing a point on a line segment. If one dowel is twice as long as another and the longer dowel is broken at random, the probability that the three pieces will form a triangle is 1/2.

5 The Computer Problem

Ms. Xavier and Mr. Young use the computer in the principal's office to generate worksheets for their mathematics classes. Every afternoon both teachers go to the principal's office between 4:00 and 5:00 p.m. to run their programs. Each program requires 10 minutes of computer time. No one else at the school ever uses the computer during this time, so Xavier and Young usually do not have to wait to get on the computer. But sometimes one of them is already on the computer when the other arrives to use it. Assuming that both teachers arrive at random times between 4:00 and 5:00, what is the probability that one of them will have to wait to get on the computer?

Someone will have to wait if Ms. Xavier and Mr. Young arrive at the principal's office within 10 minutes of each other. For instance, suppose Ms. Xavier arrives at 4:13. If Mr. Young arrives at 4:20, then he will have to wait. To avoid waiting, he would have to arrive later than 4:23. On the other hand, if Ms. Xavier arrives at 4:13 and Mr. Young arrives at 4:05, then she will have to wait. To avoid her waiting, Mr. Young would have to arrive before 4:03.

The geometric aspects of this problem are not obvious. If it did not appear in a unit entitled *Geometric Probability*, you would not suspect that its solution had anything to do with geometry. How can we use geometry to model this problem? We need to identify what events are occurring randomly, and then establish a model in which these events are represented by points. The random events are the times when the teachers arrive at the principal's office. We can make these arrival times correspond to points by counting the number of minutes after 4:00 that each person arrives at the principal's office. Let x be the number of minutes for Ms. Xavier and y be the number of minutes for Mr. Young. Then the point (11, 55) means that Ms. Xavier arrived at 11 minutes after 4:00 and Mr. Young arrived at 55 minutes after 4:00. These times are not within 10 minutes of each other, so Ms. Xavier was finished with the computer before Mr. Young arrived, and he did not have to wait. (Note that Mr. Young's program will finish at 5:05; this does not violate the conditions of the problem, which specify only that *arrival* times are between 4:00 and 5:00.)

Where is the sample space? Clearly, x and y can both have any value from 0 to 60. This means that the sample space is a 60×60 square in the coordinate plane. Its area is 3600 (see Figure 10).

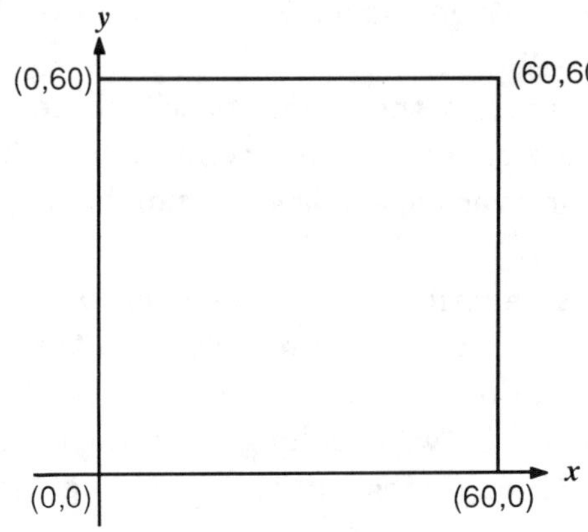

Figure 10: Sample Space for Computer Problem

Where is the event space? To answer this question, we need to think about what kinds of values for x and y will result in a teacher having to wait. One way to proceed is to test individual points within the sample space. For instance, the points $(0, 10)$ and $(60, 50)$ are on the boundary of the event space. Another way to proceed is to use algebra. It should be obvious that if $y = x$ then someone will have to wait. By how much can x and y differ and still cause someone to wait? Suppose Ms. Xavier arrives at x minutes after 4:00. Then Mr. Young will have to wait if his arrival time is within 10 minutes; that is, if $y < x + 10$. Similarly, Ms. Xavier would have to wait if Mr. Young arrived less than 10 minutes before she did; that is, if $x - 10 < y$. So one of the teachers has to wait if the values of x and y satisfy the inequality

$$x - 10 < y < x + 10.$$

The points (x, y) in the event space are shaded in Figure 11; these are the points that satisfy the inequality. You can determine the area of this wedge-shaped region by subtracting the areas of the two triangles from the area of the square. The wedge's area is 1100, the square's area is 3600, and the probability of someone having to wait is $11/36$, or 0.31.

Before proceeding, take some time to reflect on the process involved in solving the computer problem. Make sure you understand the relationship between the problem itself and the geometry used to model the problem. The real significance of the problem is in the way the arrival times are represented by points rather than in the answer to the question about probability.

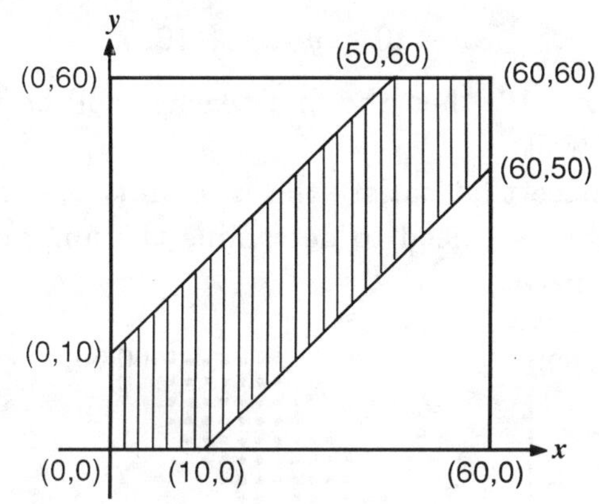

Figure 11: Event Space for Computer Problem

6 The Computer Problem Revisited

You can really appreciate the power of the model we created for the computer problem if you consider the following variation. Suppose Ms. Xavier and Mr. Young always stop and talk with the secretary whenever they go to the principal's office. They continue talking until they hear the tick of the clock on the wall; then they go in and use the computer. The effect is that their arrival times can no longer take on any value between 4:00 and 5:00 but are restricted to integer minutes: 4:00, 4:01, 4:02, ..., 4:58, 4:59, or 5:00. Under these conditions, what is the probability that one of them will have to wait to use the computer?

If x and y still represent the number of minutes after 4:00 that they start their programs, then x and y are both members of the set $\{0, 1, 2, \ldots, 58, 59, 60\}$. Under these conditions, the sample space is no longer a 60×60 square but is instead a 61×61 square grid of points (see Figure 12). The point (20.4, 32.7) is not in the sample space, since the arrival times are limited to integer minutes.

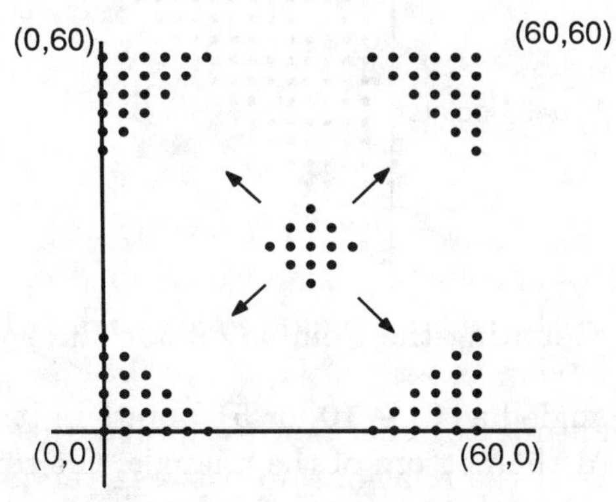

Figure 12: New Computer Problem

Within this grid, the event space consists of points (x, y) whose coordinates satisfy the inequality

$$x - 10 < y < x + 10.$$

Note that if $y = x + 10$ or $y = x - 10$, then the first teacher will be finished before the second one arrives and no one will have to wait.

The points in the sample space that satisfy the inequality $x - 10 < y < x + 10$ are shown in Figure 13. To find the probability, we need to determine the number of points in the event space and the number in the sample space.

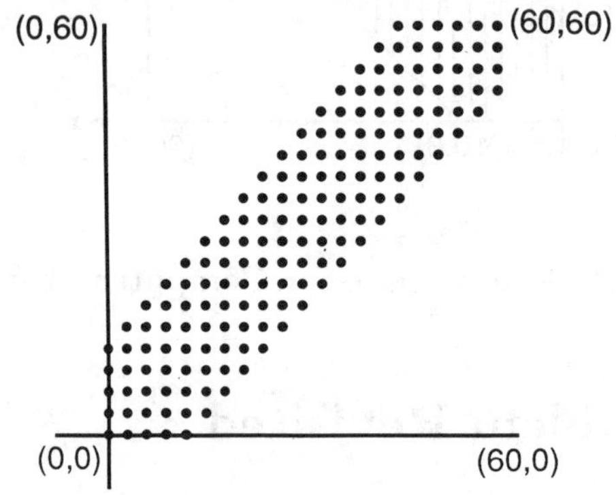

Figure 13: Event Space for New Computer Problem

The sample space has 61 points on each side, so it contains 61 × 61, or 3721, points in all. An efficient way to find the number of points in the event space is to count the points in each of the two triangles that are outside of the event space, and then subtract this number from 3721. Let's focus on the triangle that is in the upper left; we will develop a way to determine the number of points in this triangle without having to count them (see Figure 14).

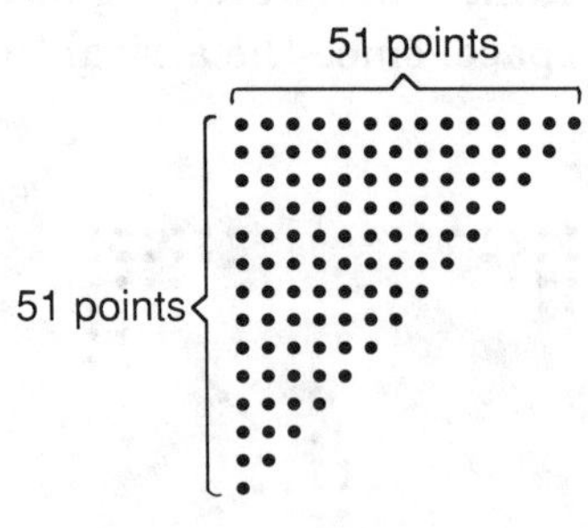

Figure 14: Counting the Points Outside the Event Space

The topmost row of this triangle has 61 − 10, or 51 points in it. The second row has 50 points, the third has 49, and so forth. At the bottom of the triangle, the rows have 3 points, then 2 points, and then 1 point, respectively. The total number of points is equal to

$$51 + 50 + 49 + \ldots + 3 + 2 + 1.$$

An efficient way of adding this list of numbers is shown below. Let the sum of the numbers be S.

$$\begin{array}{ccccccccccccccc} S & = & 51 & + & 50 & + & 49 & + & \ldots & + & 3 & + & 2 & + & 1 \\ S & = & 1 & + & 2 & + & 3 & + & \ldots & + & 4 & + & 50 & + & 51 \end{array}$$

Adding yields

$$2S = 52 + 52 + 52 + \ldots + 52 + 52 + 52.$$

There are fifty-one 52's, thus

$$2S = 52(51) = 2652;$$

$$S = 1326.$$

This means that there are 1326 points in each of the two triangles. Since the two triangles are congruent, there are $2(1326) = 2652$ points outside of the event space. Therefore, there are $3721 - 2652 = 1069$ points inside the event space. The probability that one of the teachers will have to wait is equal to 1069/3721, or about 0.29.

In this version of the computer problem, the restriction of the arrival times to integers has important implications for the solution of the problem. *The sample space and the event space both contain a finite number of points; to find the probability we count points instead of computing areas.* Another consequence of the integer arrival times is that the boundary makes a difference; in particular, the event space consists of solutions of $x - 10 < y < x + 10$, and the probability would be different if we considered $x - 10 \leq y \leq x + 10$. The probability that $y - x$ is exactly equal to 10 or -10 is not zero.

So far we have looked at two different versions of the computer problem. The case with integer arrival times had a sample space that contained a finite number of points; mathematicians would call this the *discrete* case. The case with unrestricted arrival times had a sample space that contained an infinite number of points; this is the *continuous* case. In the discrete case we counted points, and in the continuous case we computed areas. In both models, events are modeled by points and the random nature of the events is represented by a random choice of points from the sample space. Note that the discrete case could be solved by using traditional counting probability, but it turns out to be easier to use the model we created for the continuous case.

6.1 Exercises

1. In the continuous case, the probability that $y = x + 10$ or $y = x - 10$ is zero. In the discrete case, the probability that $y = x + 10$ or $y = x - 10$ is not zero. Verify that in the discrete model the probability that either $y = x + 10$ or $y = x - 10$ is 102/3721.

2. Rework the computer problem under the condition that arrival times can only be multiples of 5 (i.e. 4:00, 4:05, 4:10, ..., 4:55, 5:00). What do you expect to happen to the probability under this condition? Find the new probability.

3. Suppose Ms. Xavier and Mr. Young arrive at the principal's office at random times between 8:00 a.m. and 4:00 p.m. The schedule of classes is such that arrival times are only on the quarter hour: 8:00, 8:15, 8:30, ..., 3:30, 3:45, 4:00. Under this condition, find the probability that a teacher will have to wait to use the computer.

7 The Random Sum Problem

Try the following simple experiment. Use a computer or calculator to choose two random numbers between 0 and 5, exclusive. What kinds of numbers do you get for the sum? It should be clear that the sum can take on values between 0 and 10. We are interested in studying how the sums are distributed.

Table 15 displays forty random sums.

3.13388	5.11567	8.87824	3.12656	4.86911
6.32489	2.26661	5.41372	4.62745	6.31414
5.48868	6.25961	8.94927	4.52895	8.40610
1.72441	6.10046	4.00064	3.63134	6.15140
6.09184	4.95964	4.98476	1.71774	2.57381
5.32376	2.82664	4.86021	1.60581	3.06145
6.26421	9.57602	7.30085	5.43012	2.23365
2.59858	4.34509	5.08941	1.45722	5.44831

Table 15: Forty Random Sums

There are so many digits in the table that it is difficult to make sense of what is going on. Let's look at just the integer part of each sum. In Table 16, the sums have been truncated; a 3 represents a number between 3 and 4. Study the entries and notice what kinds of numbers appear with the greatest frequency.

3	5	8	3	4
6	2	5	4	6
5	6	8	4	8
1	6	4	3	6
6	4	4	1	2
5	2	4	1	3
6	9	7	5	2
2	4	5	1	5

Table 16: Integer Parts of Same Random Sums

The software that has been provided will generate more random sums. When the software is run, how many sums do you think will be less than or equal to 5? Do you think that 60% of them will be between 0 and 6? To analyze the distribution of the sums, we will answer the following questions:

1. What is the probability that the sum is less than or equal to 3?

2. What is the probability that the sum is less than or equal to 8?

3. What is the probability that the sum is greater than or equal to 6?

We will use a geometric model to find these probabilities. By now you should be developing a sense of how this problem can be modeled using geometry. The use of two numbers should make you think of the coordinate plane. Randomly choosing the numbers from between 0 and 5 is equivalent to randomly choosing a point (x, y) from a square whose corners are $(0,0), (0,5), (5,0)$, and $(5,5)$. This model will enable us to answer the probability questions we posed earlier.

1. What is the probability that the sum is less than or equal to 3? For this event, the event space consists of points on or below the line $x + y = 3$. This region is a triangle with base 3, height 3, and area 4.5 (see Figure 17). Since the area of the sample space is 25, the probability that the sum is less than or equal to 3 is 4.5/25, or 0.18.

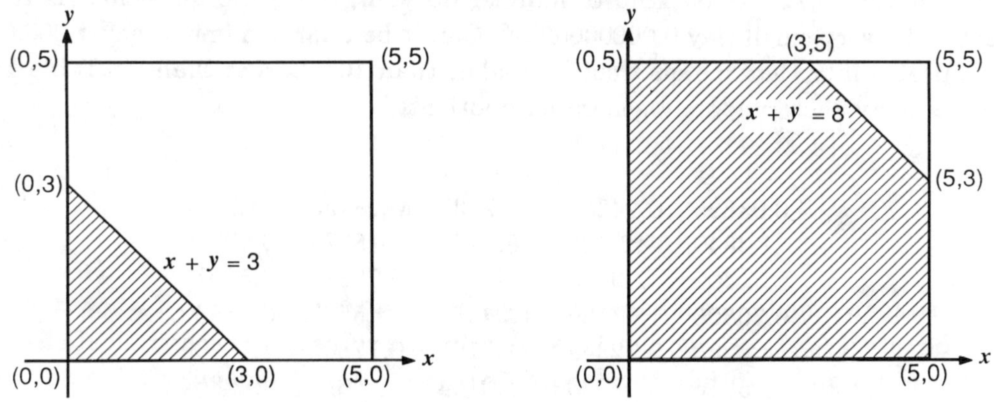

Figure 17: Event Spaces for Random Sum Problems

2. When we try to find the probability that the sum is less than or equal to 8, the event space is bounded by the line $x + y = 8$ (see Figure 17). The event space is a pentagon; its area is easily found by subtracting the area of the triangle in the upper right from the area of the square. The triangle has base 2, height 2, and area 2, so the area of the pentagon is $25 - 2$. The probability that the sum is less than or equal to 8 is 23/25, or 0.92.

3. For a sum greater than or equal to 6, the event space is bounded by the line $x + y = 6$. This gives a triangular region in the upper right-hand corner of the sample space. Its area is 8, and the required probability is 8/25, or 0.32.

Class Practice

1. What is the probability that the random sum is between 4 and 9?

2. Find the probability that the sum is less than or equal to 10.

3. Table 18 gives theoretical probabilities for the random sums. Use the geometric model to verify entries in this table.

t	1	3	5	7	9	11	13
Probability that sum is $\leq t$	.02	.18	.5	.82	.98	1	1

Table 18: Random Sum Problem

8 The Random Quotient Problem

This problem is a variation of the random sum problem. Instead of adding the two randomly chosen numbers, now we will divide them. Choose random numbers x and y from between 0 and 2 and compute the quotient y/x. What kind of number do you get for the quotient? Is it possible for the quotient to be very small, say 0.0000000001? Can it be a large number, say 1,000,000,000,000? What is the probability that the quotient is smaller than 10? Larger than 5? Table 19 shows the first 5 digits of forty such randomly generated quotients.

0.03825	0.24283	1.22533	0.28153	1.1623
1.3849	13.582	18.5175	1.8475	0.60769
25.409	2.9400	0.59791	0.72875	0.57977
0.70806	1.67126	1.24561	4.5567	0.64569
4.9283	0.72345	1.7197	0.77805	1.78606
1.5656	0.97930	1.0164	0.3615	92.864
5.1845	0.45453	0.15551	17.184	2.3429
0.96353	2.4629	3.4140	0.57341	1.8959

Table 19: Forty Random Quotients

Again, there are so many digits in the table that it is difficult to make sense of what is going on. Table 20 shows the integer part of each quotient. Study the entries and notice what kinds of numbers appear with the greatest frequency.

0	0	1	0	1
1	13	18	1	0
25	2	0	0	0
0	1	1	4	0
4	0	1	0	1
1	0	1	0	92
5	0	0	17	2
0	2	3	0	1

Table 20: Integer Parts of Same Random Quotients

The software that has been provided will generate more random quotients; it will also truncate the decimal part of the quotient. Before running the program, guess how many quotients will be less than 10. How many will be greater than 5? How will the quotients be distributed?

Were you surprised that close to half of all the quotients are between 0 and 1? How many are less than 10? Between 5 and 10? If you run the simulation several times you may be able to see "on average" how the random quotients are distributed.

Repeated simulations show a significant number of quotients between 0 and 1. How can we account for these? To analyze the distribution of the quotients, we will answer several probability questions about the quotients.

1. What is the probability that the quotient is equal to 1?

2. What is the probability that the quotient is less than 1?

3. What is the probability that the quotient is between 0.4 and 0.5?

4. What is the probability that the quotient is between 4 and 5?

We first need to set up a geometric model for the problem. Since x and y are both chosen between 0 and 2, the sample space is a 2×2 square in the coordinate plane. This model will enable us to answer the probability questions posed earlier.

1. What is the probability that the quotient is equal to 1? Since $y/x = 1$ is equivalent to $y = x$, the event space is a segment of the line $y = x$. The line has no area, and the probability is zero. (This result may make more sense if you first spend some time thinking about the distinction between continuous and discrete.)

2. What is the probability that the quotient is less than or equal to 1? In this case, the event space is the region below the line $y = x$. This is because $y/x \leq 1$ can be rewritten as $y \leq x$. It should be clear that the probability is 1/2. Do you see now why about half of the entries in Table 20 are zeros?

3. When we want to find the probability that the quotient is between 0.4 and 0.5, we need to consider the inequality $0.4 \leq y/x \leq 0.5$. This is equivalent to $0.4x \leq y \leq 0.5x$, so the event space is bounded by the lines $y = 0.4x$ and $y = 0.5x$. This region is a triangle with height 2 and base 0.2. Its area is 0.2, so the probability is 0.2/4, or 0.05 (see Figure 21).

4. For a quotient between 4 and 5, the event space is again a triangle; this time it is bounded by the lines $y = 4x$ and $y = 5x$. Its base is 0.1, its height is 2, and its area is 0.1. Therefore, the probability is 0.1/4, or 0.025 (see Figure 21).

The answers to these four questions should help show how the probability that the quotient is less than or equal to t depends on t.

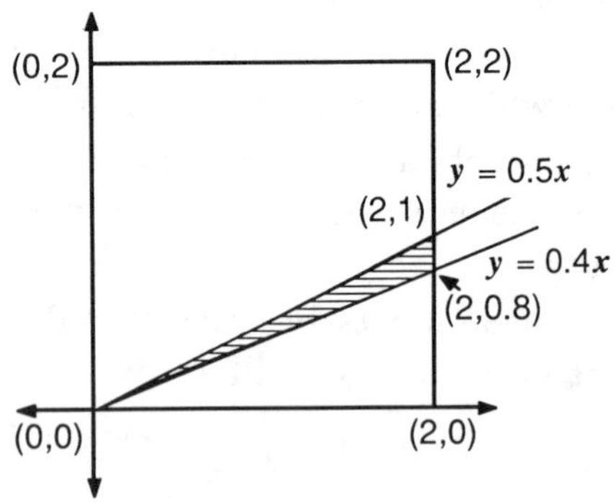

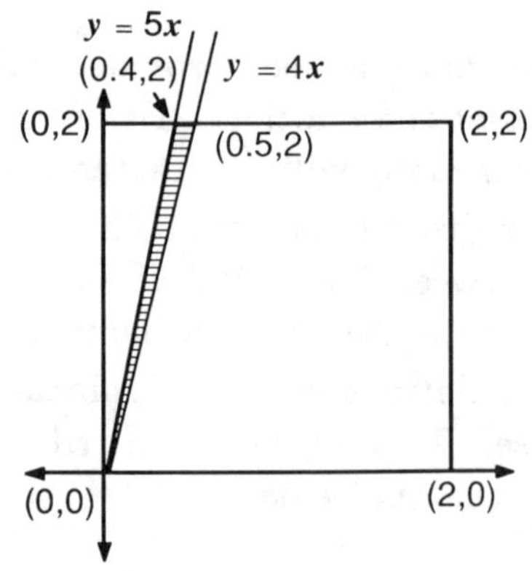

Figure 21: Event Spaces for Random Quotients

8.1 Exercises

1. What is the theoretical probability that the random quotient of numbers between 0 and 2 is smaller than 10? How does this compare with the empirical result from the computer simulation?

2. What is the probability that the random quotient is greater than 5?

3. Find the probability that the quotient is between 5 and 10.

4. For values of t that are less than 1, what is the probability that the random quotient is less than or equal to t? (Give your answer in terms of t.)

5. For values of t that are greater than 1, find the probability that the quotient is less than or equal to t. (Give your answer in terms of t.)

6. Find the probability that the quotient is between 0.0 and 0.1. Try between 0.1 and 0.2. What about between 0.3 and 0.4? Can you predict the probability that the quotient is between 0.7 and 0.8? Make a general conclusion about the distribution of quotients that are less than 1.

7. Find the probability that the quotient is between 1.1 and 1.2. What about between 1.3 and 1.4? Can you predict the probability that the quotient will be between 1.7 and 1.8? What general conclusion can you make about the distribution of quotients that are greater than 1?

8. Find the probability that the first ***nonzero*** digit in the quotient is 7. HINT: The probability is not 1/9 because the probabilities are not uniformly distributed.

9. Two real numbers are chosen at random from between 0 and 10. Find the probability that the product of the numbers is less than 25. HINT: The graph of the equation $xy = 25$ is a hyperbola that has the x- and y-axes as asymptotes. You will need to estimate the area of the event space.

9 Some Thoughts about Models and Mathematics

When children think about models, they are generally considering some kind of a toy, perhaps an airplane or a dinosaur. When scientists and mathematicians think about models, they are generally considering a model as a tool, even though they may be thinking about the same airplane or dinosaur. Scientists and mathematicians use models to help them study and understand the physical world. People in all walks of life use models to help them solve problems; problems in this course will involve models used by bankers, city planners, anthropologists, geologists, economists, automobile makers, and many, many others.

So just what is a model? Models are representations of phenomena. In order to be useful, a model must share important characteristics with the phenomenon it represents, and it must also be simpler than what it represents. A model usually differs significantly from what it represents, but these differences are offset by the advantage that comes from simplifying the phenomenon. A good example is a road map, which models the streets and highways in a particular area. Clearly, a map has a lot in common with the actual streets and highways — it shows how roads are oriented and where they intersect. A road map simplifies the situation; it ignores stoplights, steep ascents, and back alleys and instead focuses on major thoroughfares. Such a map is very useful for traveling from Raleigh to Rocky Mount but is not much good for finding the quickest route to the Burger King or the best street for skateboarding. Road maps, and most other models, are useful precisely because they ignore some extraneous information and thereby allow you to see other information more clearly. Another fairly common model is an EKG, which models the electrical activity of the heart. The EKG is an excellent model when used to determine the heart rate or to find which regions of the heart may be damaged after a heart attack. It is a very poor model for determining the volume of blood flowing through the heart or the condition of the valves. Different models emphasize different aspects of a phenomenon; the choice of what model to use depends on what aspect is under investigation.

The ability to predict is the ultimate test for a model. A good model allows us to make accurate predictions about what will occur under certain conditions. If what actually occurs is very different from our prediction, then the model is of little use.

Scientists and mathematicians often need to update or revise models as more is learned about the phenomenon under study. Sometimes a model needs to be completely discarded and replaced with a new one. The pre-Columbian model of the flat world was discarded, and Ptolemy's geocentric model of the universe was revised by Copernicus and then overthrown by Galileo.

Even though Issac Newton's models for the actions of a gravitational field have been replaced by Einstein's relativistic model, we still use Newtonian physics under everyday conditions because it is easier and because it gives reasonably accurate results. The aspects of Einsteinian mechanics that are ignored are largely irrelevant in most everyday applications, so the Newtonian model is still a good one.

As we move through the course, we will encounter phenomena that we want to know more about. Our task will be to find a mathematical expression or formula or picture that mimics the phenomenon we are interested in. This model must accurately represent the aspects of the phenomenon that we care about, but it may be very different from the phenomenon in other ways. For instance, in the triangle problem we use a line segment to represent a dowel. The salient feature of the dowel is its length, and a line segment can accurately represent its length. Clearly, a line segment represents neither the dowel's diameter nor the wood from which it is made, but these characteristics are not important aspects of the problem we are trying to solve. The fact that the line segment does not represent these aspects of the dowel does not diminish its usefulness for this problem.

To be able to find a model to represent a problem, we need to have a large tool kit of mathematical information and techniques at our disposal. The fundamental concepts learned in Algebra 1, Geometry, and Algebra 2 are all a part of our tool kit. We will also use the computer as a tool to simulate the phenomenon we are studying. Probably the most important tool necessary for model making is an inquisitive mind and a determined spirit.

Often we will not stop after we have developed one model but will form two or three to get a better view of the subject. For example, suppose a rock is thrown into the air. How can its path be modeled? It would be informative if we had a picture of the flight. We can use a graph as a model for this phenomenon (see Figure 22).

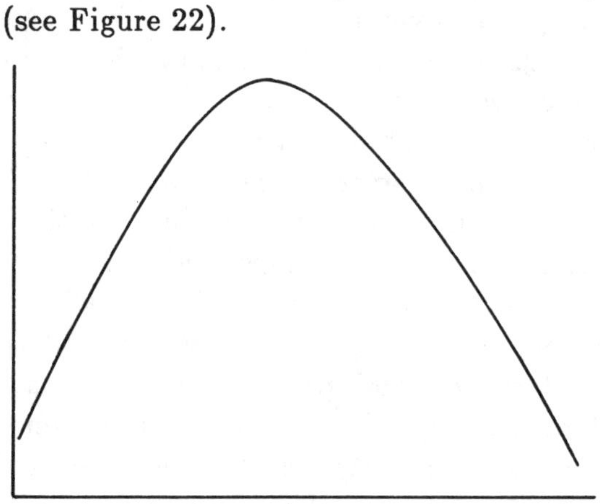

Figure 22: Model 1

We could also describe the flight by an equation. If the height above the ground is called h and the horizontal distance away from the thrower is called d, then we can represent the flight by the equation

$$\text{Model 2:} \quad h = -\frac{2}{5}d^2 + \frac{4}{5}d + 1$$

We could also introduce a third variable to represent time. Then we could describe the path by the system of equations that are called parametric equations and two separate graphs, one showing d and t and the other showing h and t (see Figure 23). (We will study parametric equations later in this course.)

$$\text{Model 3:} \quad d = 5t \qquad\qquad \text{Model 4:} \quad h = -10t^2 + 4t + 1$$

Notice that these models do not give complete information about the problem. We cannot tell from the models what type of rock was thrown, who threw it, or why. These aspects of the problem are not relevant — all we really care about is distance and time.

Which model of the rock throwing gives the most information? Well, that depends on what question we are asking. If we are just interested in the height of the rock, then models 4 and 6

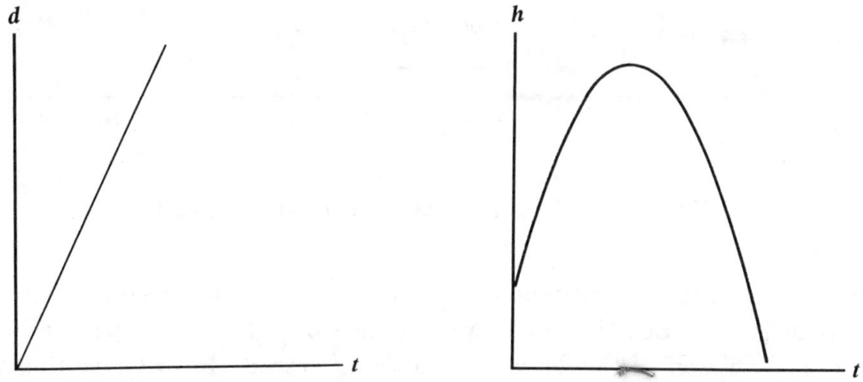

Figure 23: Models 5 and 6

give us all the information we need. If we would like to know the path that the rock followed, then models 1 and 2 should be used. The strength of the model is in what it emphasizes and what it ignores.

The process of developing a model that is appropriate for a given situation does not follow any prescribed set of rules. It is often largely a matter of common sense, together with a certain amount of trial and error. Model making is one of the creative aspects of mathematics, and proficiency at this skill comes with experience. To illustrate the process of creating a model, we present the following example.

9.1 The Tape Recorder Problem

A tape recording is made of a meeting between two businessmen. Their conversation starts at the 21st minute on the tape, and lasts only 8 minutes. The tape will record for 60 minutes. While playing back the tape one of the businessmen accidentally erases 15 minutes of the tape, but he doesn't know where.

1. Find the probability that the entire conversation was erased.

2. Find the probability that some part of the conversation was erased.

3. Suppose the exact position of the conversation on the tape is not known, except that it began sometime after the 21st minute. Find the probability that the entire conversation was erased.

In part 1, a single random event occurs — that is, the starting time of the 15-minute erasure. If we let x be the number of minutes from the beginning of the tape to the start of the erasure, then x can vary from 0 to 45. (Do you see why x cannot take on values between 45 and 60?) Using a number line to represent the tape, we can see that the sample space is 45 units long. An illustration is shown in Figure 24.

Where does the event occur? The conversation lasts from the 21st minute to the 29th minute, so the erasure can start as early as the 14th minute and as late as the 21st minute in order to erase the entire conversation. Thus, the value of x must be between 14 and 21 in order to be in the event space. The event space is 7 units long, the sample space is 45 units long, and the probability is 7/45, or about 0.16.

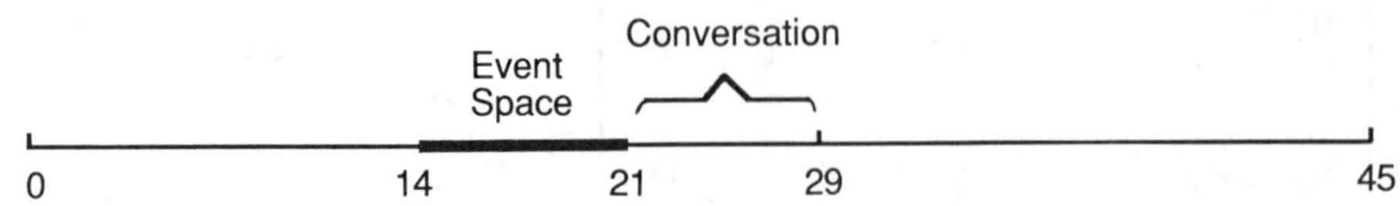

Figure 24: Tape Recorder Problem – Part 1

For part 2, we can still let x represent the starting time of the erasure. The sample space is the same as it was in part 1, but the event space is larger. The event space now includes values of x that result in any portion of the conversation being erased. In order to find the event space, the situation can be analyzed algebraically, or you can use trial and error. You should find that x can be anywhere between 6 minutes and 29 minutes. So the event space is 23 units long, and the probability is 23/45, or about 0.51.

In part 3, two random events occur — one is the starting time of the erasure and the second is the starting time of the conversation. Let x represent the starting time of the erasure, and let y be the starting time of the conversation. Since the erasure is 15 minutes long, x can vary from 0 to 45. Since the conversation is 8 minutes long, y can vary from 21 to 52. The sample space consists of all pairs (x, y) in the rectangle in Figure 25.

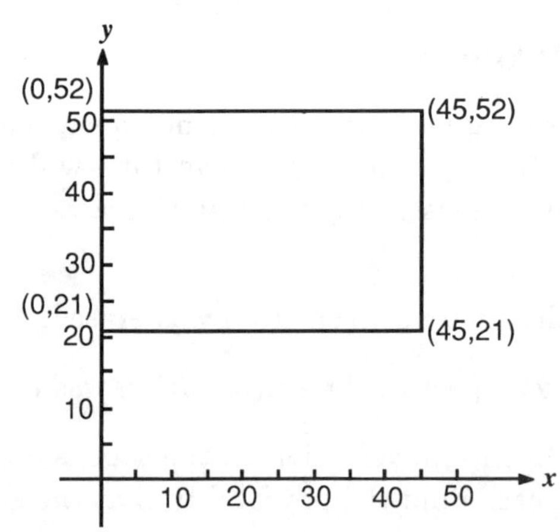

Figure 25: Sample Space for Part 3 of Tape Recorder Problem

Which points in the sample space are in the event space? A good way to begin to find the event space is by testing individual points. Consider the point (15, 30). This corresponds to the erasure starting at the 15th minute of the tape and the conversation starting at the 30th minute. The erasure would be over before the conversation began, so this point is not in the event space. By similar reasoning, you should be able to decide that (17, 23) and (20, 27) are in the event space, whereas (30, 25) is not. In general, it is clear that the conversation must not start before the erasure — thus, $y \not< x$, or $y \geq x$. Also, the conversation must start by the time 7 minutes of the erasure have gone by, so that there will still be 8 minutes of erasure left to erase the 8-minute conversation — thus, $y \leq x + 7$. This means that a point is in the event space if its coordinates satisfy the inequality $x \leq y \leq x + 7$. The event space is shaded in Figure 26. Its area is 192.5, the area of the sample space is 1395, and the probability is 192.5/1395, or about 0.14.

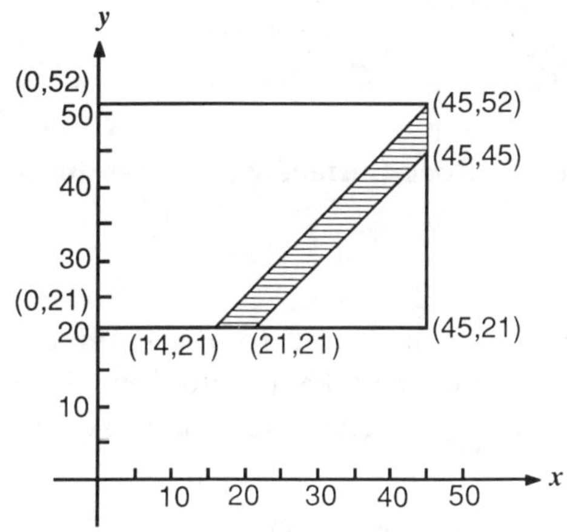

Figure 26: Event Space for Part 3 of Tape Recorder Problem

Remember that the real significance of the tape recorder problem is the model we used to represent it. The areas of the regions and the actual probabilities are not really important — what you should appreciate about the problem is the model.

10 Applications to Geology

Geologists use a method based on the principles of geometric probability to determine the composition of rocks that contain several different minerals. The rock in question is sliced and polished, and then parallel lines are inscribed on the surface. Geologists measure the total length of the inscribed line segments.

Suppose that $x\%$ of the rock were composed of mineral A. We would then expect that $x\%$ of the total length of the inscribed lines would lie through mineral A. Now if a geologist finds experimentally that $x\%$ of the length of the line falls on mineral A, she can reasonably conclude that $x\%$ of the rock is composed of mineral A. (See Figure 27.)

An alternative method of determining rock composition involves counting points rather than measuring lengths. The surface of a rock slice is overlaid with a regular grid pattern of dots. The total number of dots over the rock is the sample space and the number of dots over each mineral is that mineral's event space.

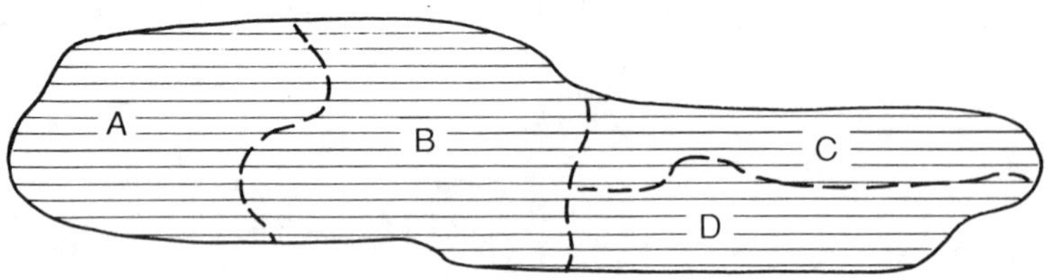

Figure 27: Rock Surface with Lines Inscribed

10.1 Exercises

1. Students working on a biology project need to approximate the surface area of Clear Lake. They have an aerial photograph of the lake, pictured in Figure 28. Discuss how they can use principles of geometric probability to determine the area.

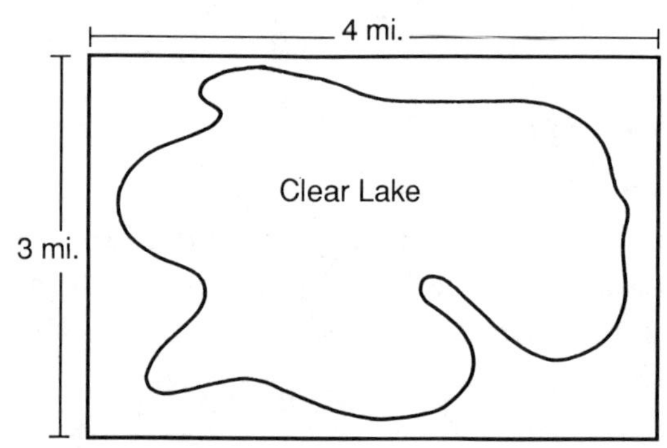

Figure 28: Clear Lake

2. Figure 29 shows the cross section of a hypothetical rock that is composed of minerals A, B, C, and D.

   (a) Use the method of parallel lines to determine what percent of the rock is mineral A, what percent is mineral B, and so forth.

   (b) Use the method of dots to determine the composition of the rock.

   (c) Use your knowledge of geometry and areas of geometric shapes to determine what percent of the total area is represented by each mineral. Use these results to help you judge the accuracy of the methods you used in the first two parts.

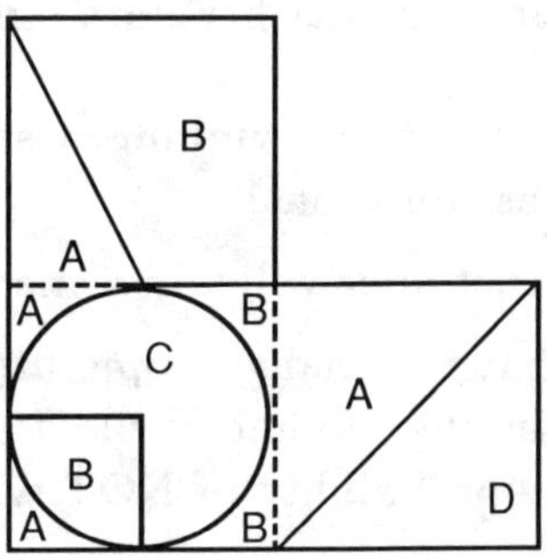

Figure 29: Rock Containing Minerals A, B, C, and D

11 Exercises

1. Suppose 100 darts are thrown at random at a dart board and 26 of the darts land in a region that is painted red. Make a sketch to show what the dart board might look like and how much of it is painted red.

2. An electric power line is suspended between two poles that are 1200 feet apart. High winds result in a break at a random point on the line. What is the probability that the break is less than 30 feet from a pole?

3. Suzy's Subway Shop has a square dart board that measures 18 inches × 18 inches hanging on the wall. Customers can win free sandwiches by throwing darts at the board. The dart board contains three concentric circles with their centers at the center of the square. The radii of the circles are 3 inches, 4 inches, and 6 inches, respectively. The prize for throwing a dart into the inner circle (the bull's eye) is a free large sandwich. Customers win a medium sandwich if their dart lands in the circle of radius 4 (excluding the bull's eye). The prize for throwing a dart inside the circle of radius 6 (excluding the two inner circles) is a free small sandwich. A customer wins nothing if his or her dart lands outside the largest circle.

   (a) If you want to find the probability of winning a sandwich, you need to make two assumptions. What are those assumptions?

   (b) Find the probability that a customer will win a medium sandwich.

   (c) The management decided that too many people are winning sandwiches, so they want to change the chances of winning. What would the radius of the largest of the three circles have to be so that the probability of NOT winning a sandwich would be 0.95?

4. Refer to the parachute problem. Recall that the square field measures 2 km on each side. Suppose the parachutist gets caught if she lands within x km of a tree.

   (a) Find x so that the probability of getting caught is 0.01. What value of x gives a probability of 0.10?

   (b) For what value of x will the probability of getting caught be 1? (The answer is not $2/\sqrt{\pi}$. With this radius, there is still a possibility of not getting caught because the event space does not cover the entire sample space.)

   (c) Suppose $x = 2/\sqrt{3}$. Find the probability that she will get caught in a tree. Figure 30 should help you think about how to approach this problem; the quarter circles in each corner of the field overlap.

5. A surveillance company has detection devices that illuminate a region shaped like a quarter circle with radius of 100 yards. These devices will be distributed along a border pass that is 1000 yards wide (see Figure 31.)

   (a) Suppose 5 of the devices are used to survey the pass. If an intruder enters the pass at a random point, what is the probability that he will be detected?

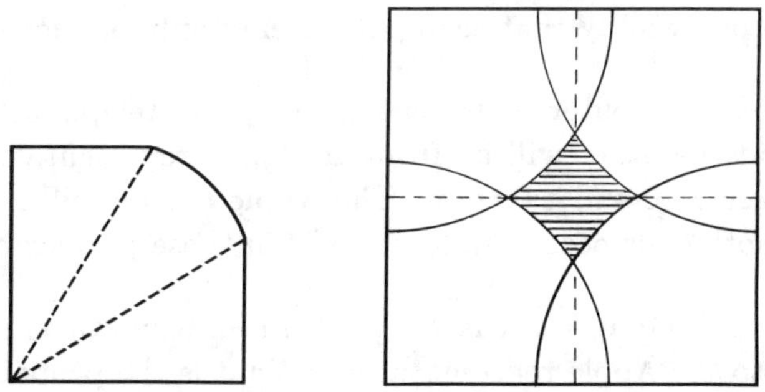

Figure 30: Parachute Problem

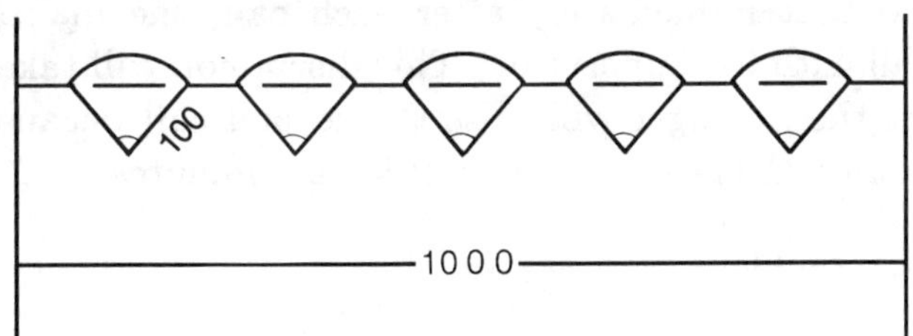

Figure 31: Surveillance Problem

(b) How many devices would be needed to ensure that the probability of detecting an intruder would be greater than 0.9?

6. Refer to the triangle problem.

   (a) Suppose that one dowel is three times as long as the other. What is the probability that a triangle will be formed when the longer dowel is broken at a random point?

   (b) Suppose that one dowel is four times as long as the other. What is the probability that a triangle will be formed?

   (c) If one dowel is n times as long as the other, what is the probability that a triangle will be formed?

   (d) Suppose one dowel is twice as long as the other. Find the probability that an acute triangle is formed. A triangle with sides $a, b,$ and c is acute if and only if

$$a^2 + b^2 > c^2$$

$$\text{and} \quad a^2 + c^2 > b^2$$

$$\text{and} \quad b^2 + c^2 > a^2.$$

7. Max and Yvonne want to meet at the ice-cream shop to enjoy their favorite after-dinner treat. Both agree to arrive sometime between 8:00 and 8:30; they also agree that the first person to arrive will buy two cones, and then wait for the other person. If the second person has not arrived within 12 minutes, the first person will start to eat that person's cone and will then

continue to wait for the friend. Of course, if they arrive at the same time, each will buy his own cone. Find the probability that each person eats only one ice-cream cone.

8. The manager of a busy computer center is expecting two telephone calls; one will be from an IBM salesperson and the other will be from an Apple representative. The IBM person will call sometime between 2:00 and 4:00 p.m. The Apple person will call sometime between 2:30 and 3:15. Assume both calls occur randomly within these periods of time.

   (a) Suppose that the manager needs to get some information from the IBM salesperson before talking to the Apple representative. What is the probability that she will be able to do this (i.e., that the IBM salesperson will call first)? Assume that the manager has two phone lines.

   (b) Suppose instead that immediately after each call, the manager needs to discuss the results of the call with her supervisor. This discussion will take 10 minutes. What is the probability that the manager will miss the second call because she is talking with her supervisor? Assume that each phone call lasts 5 minutes.

9. Refer to the computer problem.

   (a) Rework the problem under the condition that the teachers arrive at random times between 4:00 and 6:00. Assume that both teachers' programs require 10 minutes of computer time. Do you expect the probability of someone having to wait to increase or decrease? Find the new probability.

   (b) Rework the problem under the condition that Ms. Xavier's program takes 10 minutes to run and Mr. Young's program takes 15 min. Assume that both teachers arrive between 4:00 and 5:00. Sketch the sample and event spaces. What is the probability that one of them will have to wait?

   (c) If both teachers arrive between 4:00 and 5:00 and their programs require equal amounts of time, how long could the programs last so that the probability of someone waiting would be less than 0.2?

   (d) Suppose that the teachers arrive between k o'clock and 5:00 and their programs both require 10 minutes of computer time. For what values of k will the probability of someone waiting be less than 0.05?

10. Refer to the fairground problem. Recall that the squares are S units on a side, that the coin has a radius of R units, and that the probability of winning is given by

$$P = \frac{(S - 2R)^2}{S^2}.$$

   (a) Choose a particular value for R, say $R = 10$. Let the value of S vary from 20 to 200 in steps of 20. Make a table of values showing how P depends on S (with a fixed value for R). Now plot the ordered pairs (S, P) in a coordinate plane. What kind of relationship do you see between S and P? What happens to the value of P as S continues to increase?

(b) Choose a particular value for S, say $S = 100$. Let the value of R vary from 0 to 50 in steps of 10. Make a table of values showing how P depends on R (with a fixed value for S). Now plot the ordered pairs (R, P) in a coordinate plane. What happens to P as R increases? What limitations are there on the value of R?

(c) We can rewrite the expression for the probability of winning in terms of the variable R/S.

$$\text{Probability } = (1 - 2\frac{R}{S})^2$$

Sketch a graph of this equation with probability on the vertical axis and R/S on the horizontal axis. What must be the relationship between R and S for the probability to be 1/2?

11. Recall that in the random sum problem we determined the probability that the sum is less than or equal to t for various values of t. Find a formula that expresses the probability that the sum is less than or equal to t in terms of t. HINT: Your formula should have two parts to it — one part for t-values that are less than or equal to 5 and another part for t-values that are greater than 5.

12. Suppose two numbers are chosen at random from between 0 and 4 and then are added. For what value of k will 50% of the sums be less than or equal to k? For what value(s) of k will 100% of the sums be less than or equal to k?

12 Supplementary Exercises

1. The finale of an air show involves a parachutist who jumps from a plane and tries to land in a 10 m $\times$ 10 m pool. The pool is located within a circle that is painted bright red so that the parachutist can see it clearly from the plane before he jumps. What must be the radius of the circle so that the parachutist will have a 60% chance of missing the pool? (Assume that the parachutist has a 100% chance of landing within the red circle.)

2. A small plane left the National Airport at 12:00 for a flight that should have lasted 80 minutes. Unfortunately, the plane crashed before reaching its destination.

   (a) Find the probability that the crash occurred more than 30 minutes after its take-off.

   (b) Authorities who are investigating the crash learn that the pilot made radio contact with a control tower at 12:15. Given this information, find the probability that the crash occurred more than 30 minutes after take-off.

3. A bowling pin that is 4 inches wide is placed at the end of a bowling lane that is 42 inches in width. A bowling ball 8 inches in diameter is thrown towards the pin. Assume that the lane has no gutters, that neither the ball nor the pin can extend beyond the edges of the lane, that the ball is thrown randomly within the lane, and that the pin will fall over if it is touched by the ball.

   (a) Suppose the pin is placed in the the center of the lane. Find the probability that it will be knocked over by a randomly thrown bowling ball.

   (b) Suppose the pin is placed at the leftmost edge of the lane. What is the probability that it will be knocked over by a randomly thrown ball?

   (c) Suppose that the pin is placed randomly. If the ball is randomly thrown, what is the probability that the pin will fall?

4. Suppose two real numbers are chosen at random from between 0 and 2. Find the probability that

   (a) the sum of the numbers is less than 2;

   (b) the sum of the numbers is greater than 1;

   (c) the sum of the numbers is between 1 and 2.

5. A math class watched a computer simulation of darts being thrown at a target like the one in Figure 32. One simulation shows that 25% of the darts landed in the shaded area. Is this what you would expect? Explain your answer.

6. In the equation $px - q = 0$ the value of the coefficient p is selected at random between 1 and 2 and q is selected at random between -1 and 3. Find the probability that the solution of the equation will be less than $1/2$.

7. Refer to the parachute problem; recall that the field measures 2 km $\times$ 2 km. Assuming that the parachutist will get caught if she lands within $2/\sqrt{3}$ of a tree, find the probability that she will get caught in *two* trees.

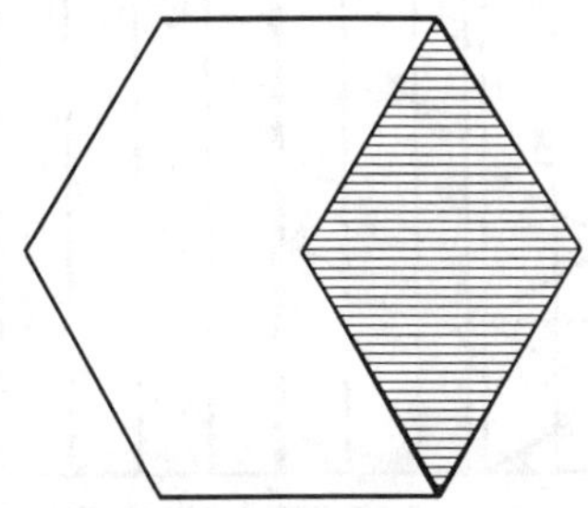

Figure 32: Dart Board Simulation

8. Contestants on a television game show spin an arrow to find out what prizes they will win. They win the grand prize if the arrow lands in the sector of the circle covered with flashing lights (see Figure 33). Students watching the show wondered how they could figure out the probability of winning the grand prize. Student A suggested working with ratios of areas. Student B wanted to compare the arc length with the circumference. Student C thought they should look at the ratio of the central angle to the total number of degrees in a circle. Discuss the strengths and weaknesses of each of the models. Which student(s) were right? Which were wrong? Why?

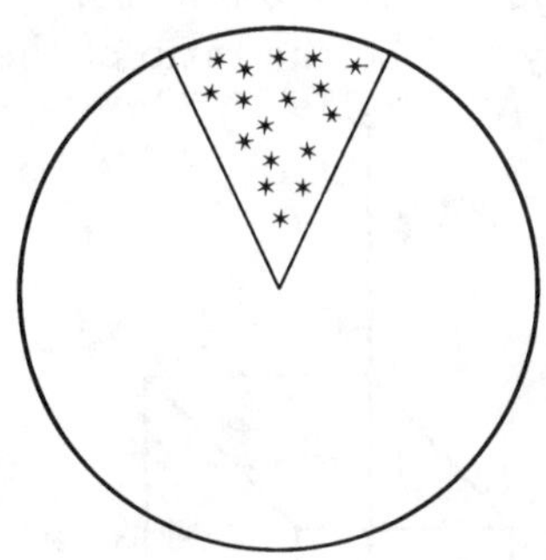

Figure 33: Game Show

9. Two lions at the city zoo inhabit adjacent cages that have a common wall. Both lions continually pace back and forth in their cages, moving from their far walls to their adjacent wall and back again (see Figure 34). Assume that the lions pace in a straight line and that they can be anywhere on this line at any given instant. Find the probability that they are at least 20 feet apart.

10. A disk of diameter A units is tossed onto a plane that is ruled with parallel lines that are spaced B units apart. Find the probability that when the disk stops moving it is crossing or touching a line. Express your answer in terms of A and B.

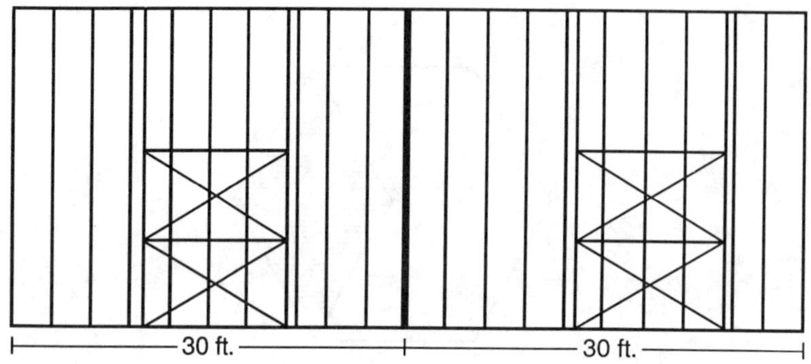

Figure 34: Lion Cage

11. A stick 1 meter long is broken at two random places in order to form three new pieces. Suppose you wanted to find the probability that the three pieces form a triangle if laid end to end. Three different models for this problem are presented below. Discuss the advantages and disadvantages of each model, and tell which model you would prefer to use to determine the probability. You do not need to actually compute any probabilities. HINT: One of the models described below is actually incorrect.

    MODEL A: Let the three pieces have lengths x, y, and z. Each piece has length between 0 and 1. The sample space is the unit cube (see Figure 35). In order to form a triangle, it must be true that $x + y > z$ and $x + z > y$ and $y + z > x$. These three inequalities define the event space. The probability will be the ratio of the volumes of the event space and the sample space.

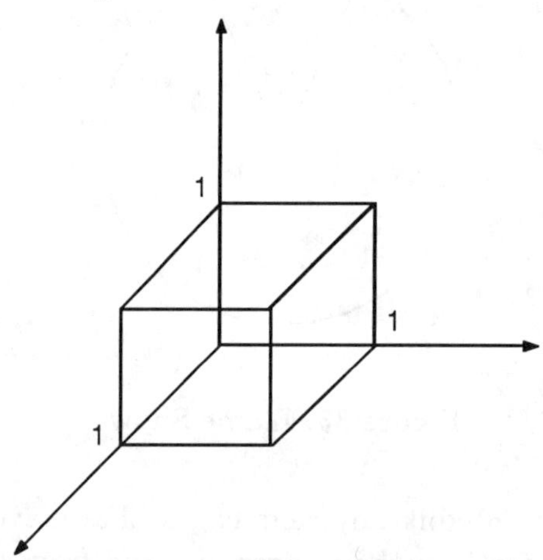

Figure 35: Sample Space for Model A

MODEL B: Let the two points where the stick is broken be x and y. Then the lengths of the three pieces are $x, y - x$, and $1 - y$. Both x and y are chosen from between 0 and 1, so the sample space is a unit square (see Figure 36). The conditions for forming a triangle are $x + (y - x) > (1 - y)$ and $x + (1 - y) > (y - x)$ and $(1 - y) + (y - x) > x$. These three inequalities define the event space. The probability will be the ratio of areas.

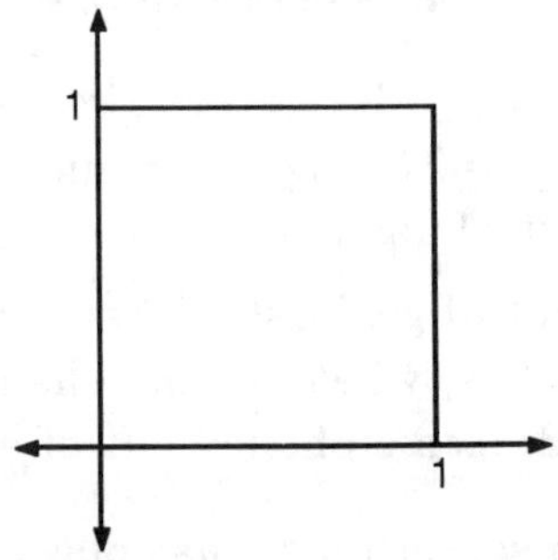

Figure 36: Sample Space for Model B

MODEL C: Let the three lengths be x, y, and $1 - x - y$. Again, x and y are chosen from 0 to 1, but it must also be true that $1 - x - y$ is between 0 and 1 — this implies that $x + y < 1$. The sample space is a triangle (see Figure 37). A triangle will be formed if $x + y > (1 - x - y)$ and $x + (1 - x - y) > y$ and $y + (1 - x - y) > x$. The probability will be the ratio of areas.

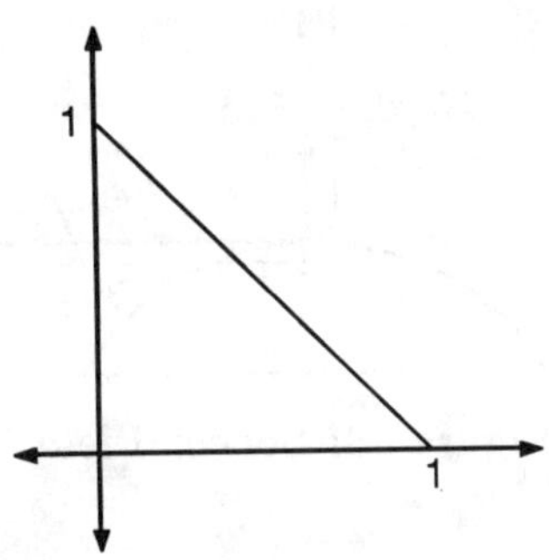

Figure 37: Sample Space for Model C

12. Refer to the previous problem. Determine the probability that a triangle is formed.

13. Josephine is standing on the train platform waiting to catch a train. A northbound train arrives at the station every 6 minutes and a southbound train arrives every 11 minutes.

    (a) Find the probability that the next train to arrive will be northbound.

    (b) Find the probability that a train headed in either direction will arrive within 3 minutes.

    HINT: Consider the number of minutes that she might have to wait.

14. Burt and Ernie decide to hold a competition to see whose car gets the better gasoline mileage. They plan to start with 12 gallons of gas in each tank and drive at the same speed until someone runs out of gas — the car that goes farthest will be the winner. Bert and Ernie

are both very competitive, and both of them sneak out the night before the race and siphon gas out of their opponent's gas tank. Assume they siphon off a random amount less than 5 gallons.

(a) Suppose Burt's car really gets 24 mpg and Ernie's gets 32 mpg. What is the probability that Burt's car will run out of gas first?

(b) Suppose Burt's car runs out of gas first. What is the probability that Ernie's car really gets better mileage? (That is, what are the chances that Burt would have run out of gas first even if no gas had been siphoned out of either tank?)

15. Sally and Sue are members of a soccer team that plays on a field that is 60 yards wide. Sally is running laterally on the 40 yard line; she is somewhere between the middle of the field and the side line. Sue is running up and down the middle of the field between the 20 yard line and the 40 yard line (see Figure 38). Find the probability that Sally and Sue are within 10 yards of each other at the instant when the goalie kicks the ball into play.

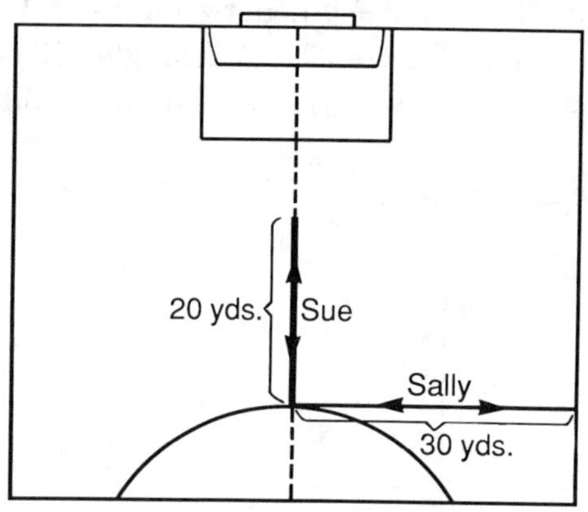

Figure 38: Soccer Field

Answers to Exercises

Class Practice from Section 1

1. Answers will vary.
2. Answers will vary; using dartboard 1 tables should show the greatest frequency for 2 or 3 darts in the shaded area, and the frequency should decrease for greater numbers of darts.
3. If you roll a pair of dice many times, on the average 6/36 of the time the sum will be seven.
4. (a) You will either win or lose. Probability gives no information about what will happen when you play once.
   (b) You expect to win 40 times.
   (c) More than \$1.50 (If x is the amount won, then long-run expected winnings are $0.4x - 0.6$. This quantity must be positive for the game to be worth playing.

Class Practice from Section 2

1. $\frac{4-\pi/81}{4} \approx 0.99$; $\frac{4-\pi/25}{4} \approx 0.97$
2. theoretical probability $= \frac{25-25\pi/4}{25}$; let empirical probability be P (obtain value from computer); set theoretical probability = empirical probability and solve for π.

Class Practice from Section 3

1. $R = 25(1 - \sqrt{.5}) \approx 7.3223$
2. guaranteed to lose if $R \geq 2S$; never guaranteed to win as long as the coin has a nonzero radius

Exercises from Section 6

1. Answers will vary.
2. probability should decrease; new probability $= 37/169 \approx 0.22$
3. $33/1089 \approx 0.03$ (Someone will have to wait only if Ms. Xavier and Mr. Young arrive at the same time.)

Class Practice from Section 7

1. 0.66

2. 1

3. Answers will vary.

Exercises from Section 8

1. 0.95

2. 0.1

3. 0.05

4. $\frac{t}{2}$

5. $1 - \frac{1}{2t}$

6. all probabilities are equal to 0.05; when a and b are both less than 1, the probability that the quotient is between a and b depends only on the difference between a and b.

7. 0.038; 0.027; can't predict; when a and b are greater than 1, the probability that the quotient is between a and b does *not* depend only on the difference between a and b, rather the probability decreases as a and b increase.

8. 0.065

9. about 0.6

Exercises from Section 10

1. Answers will vary.

2. (a) Answers will vary.
   (b) Answers will vary.
   (c) Mineral A 28%; Mineral B 35%; Mineral C 20%; Mineral D 17%

Exercises from Section 11

1. Answers will vary; about 26% of the surface should be red.

2. 0.05

3. (a) all darts hit board and darts are randomly thrown
   (b) 0.068
   (c) 2.27

4. (a) 0.113, 0.3568
   (b) $\sqrt{2}$
   (c) 0.9264

5. (a) 0.7071
   (b) 7

6. (a) $\frac{1}{3}$
   (b) $\frac{1}{4}$
   (c) $\frac{1}{n}$, 0 if $n \leq 1$
   (d) $\frac{1}{4}$

7. 0.64

8. (a) 0.4375
   (b) 0.25

9. (a) 0.16
   (b) 0.372
   (c) 6.33 min
   (d) 10:25 or earlier

10. (a) see Table 39; as S increases, so does P; P approaches 1

S	20	40	60	80	100	120	140	160	180	200
P	0	0.25	0.44	0.56	0.64	0.69	0.73	0.77	0.79	0.81

Table 39: Solution for Exercise 10(a)

(b) see Table 40; as R increases, P decreases; R must be between 0 and 50

R	0	10	20	30	40	50
P	1	0.64	0.36	0.16	0.04	0

Table 40: Solution for Exercise 10(b)

(c) $R = 0.146S$

11. for $t \leq 5 : P = 0.02t^2$; for $5 < t \leq 10 : P = \frac{25-0.5(10-t)^2}{25}$

12. $k = 4$; $k \geq 8$

Supplementary Exercises from Section 12

1. 8.92 m
2. (a) $\frac{5}{8}$
   (b) $\frac{10}{13}$
3. (a) $\frac{6}{17}$
   (b) $\frac{2}{17}$
   (c) $\frac{392}{1292}$
4. (a) $\frac{1}{2}$
   (b) $\frac{7}{8}$
   (c) $\frac{3}{8}$
5. One-third of the area of the hexagon is shaded. In the long run, we expect 33% of the darts to land in the shaded region. But 25% is not unreasonable, especially if the number of darts thrown was small.
6. $\frac{7}{16}$
7. 0.121
8. All approaches are equivalent, and all are correct. The central angle is probably easiest to work with.
9. $\frac{7}{9}$
10. $\frac{A}{B}$
11. Model C is best; easier to work with areas than with volumes, and Model B is incorrect
12. $\frac{1}{4}$
13. (a) 0.73
    (b) 0.64
14. (a) 0.89
    (b) 0.5
15. 0.13

Users's Guide for Geometric Probability Software

This software is designed to accompany the unit entitled *Geometric Probability* that has been developed by the Mathematics Department at the North Carolina School of Science and Mathematics as part of a fourth-year high school mathematics course. The software runs on an IBM PC or an IBM compatible computer with 256K memory. Although the software can be used by students on an individual basis, its best contribution to learning comes when it is used for demonstrations in front of a class.

The software is menu-driven; the purpose of this user's guide is to assist you in choosing options and interpreting screens. The utility of this guide will be maximized if it is consulted while you are actively engaged in using the software.

After you turn the computer on, insert the *Geometric Probability* disk into the the disk drive; while the disk drive light is on, the computer is loading the program. When the light goes off, you can type F1 to get more information about this and other software that accompanies the ICM course. Pressing any key will start the program and show you the main menu.

Each of the items listed on the main menu corresponds to an example that is presented in the *Geometric Probability* unit. For each item, a simulation of the experiment is performed; the user sees a visual display of the relevent random event, as well as an empirical probability. Seven different simulations are included on the diskette:

1. Dartboard Simulation
2. Parachute Simulation
3. Triangle Simulation
4. Fairground Simulation
5. Computer Time Simulation
6. Random Sums Simulation
7. Random Quotients Simulation

13 Dartboard Simulation

Key Commands A screen entitled *Key Commands* will appear at the beginning of each simulation. All of these keys can be used during the simulation as needed. The Q key enables you to quit any option at any time. This means that you do not need to wait through a simulation that you really didn't want to see. The number keys (1 through 9 at the top of the keyboard) allow you to control the speed at which the simulation is executed: the smaller the number selected, the faster the execution. Select a number any time after the simulation begins; the speed can be changed several times during the simulation. You will want to experiment to find what speed seems most appropriate for use with your students. The P key pauses the output and acts as toggle switch; pushing the button freezes the action and pushing it again unfreezes the action. This feature is often quite useful during classroom demonstrations. At the end of this screen, enter the number of times you want the event to be simulated.

Choice of Dartboard Dartboard 1 is identical to target B in Figure 1 of the *Geometric Probability* unit. Dartboard 2 is identical to the target in Figure 2.

14 Parachute Simulation

Key Commands These keys have the same effect they did in the dartboard simulation, with one addition. Pushing the space bar pauses the action and also gives some additional information about the problem, including the theoretical probability.

Tree Size The user must choose the minimum safe distance from the parachutist to the tree. The distance must be expressed as a decimal part of the side of the square. For instance, type 0.3 if the minimum safe distance is 30% of the side of the square. Note that this simulation differs slightly from the example in the unit in that it uses a 1×1 square instead of a 2×2 square.

Interpreting the Screen The display on the left shows the trees in each of the four corners, and a point represents where the parachutist lands. The display on the right shows the sample space and the event space for the geometric model of the problem.

15 Fairground Simulation

Key Commands These keys operate the same way they did in the dartboard simulation, with the addition of the space bar. Pushing the space bar pauses the action and displays additional information about the problem, including the theoretical probability.

Coin Size The user must enter the size of the coin relative to the size of the square; this ratio should be expressed as a decimal. For instance, if you want the coin's radius to be 40% of the side of the square, type 0.4.

Interpreting the Screens The left display shows where the coin lands on the board. The right display shows the sample space and event space for the geometric model.

16 Triangle Simulation

Key Commands These keys have the same effect they did in the dartboard simulation, with one addition. The user can select the number of intermediate positions for the sides of the triangle that will be shown on the screen. This selection is made with the function keys at the left of the keyboard: F1 for one intermediate position, F2 for two positions, up to F9 for nine positions. The default on this option is to show three intermediate positions. Select the number of positions while the simulation is running.

The Simulation The screen shows the two dowels laid end to end. AB is 100 units long and BD is 200 units long. An X marks the location of the break in the longer dowel. If the three resulting pieces will not form a triangle, the screen shows that no triangle can be formed. If a triangle can be formed, the two sides from the longer dowel rotate around point B and the resulting triangle is shown on the screen. The number of intermediate positions of these two sides can be controlled as described in the preceeding *Key Commands* section.

17 Computer Time Simulation

The user can choose whether or not to use the geometric version of this simulation. The nongeometric version simulates the problem by generating random arrival times at the principal's office. For teaching purposes, it is preferable to show students the nongeometric version first; it usually helps them understand the problem well enough to develop an appropriate geometric model. The geometric version of the simulation uses points to represent random arrival times. After using the nongeometric version, you must return to the main menu in order to access the geometric version.

Key Commands These commands are the same as described in the dartboard simulation. In the geometric version of this problem the space bar pauses the action and displays additional information, including the equations of boundary lines and the theoretical probability.

Choosing Parameters The user must enter starting and ending times. These correspond to the earliest and latest possible arrival times, and should be based on a 24-hour clock. For instance, if the starting and ending times are 8 a.m. and 4 p.m., type 8:00 and 16:00. The time required by each teacher to run his or her program should be entered in minutes; the duration of each program must be less than the elapsed time between starting and ending times.

Interpreting the Screens The nongeometric version displays random arrival times at the bottom of the screen. The top of the screen summarizes the conditions of the problem and gives an empirical probability.

Students may notice that the random arrival times generated by the software are all reported to the nearest minute. Although this may give the impression that the software is modeling a discrete problem, students should bear in mind that the problem is actually meant to be continuous — arrivals can occur at any time between the specified starting and ending times.

The geometric version of the software displays the sample and event spaces for the geometric model of the problem; it also gives an empirical probability.

18 Random Sums Simulation

The user can choose whether or not to use the geometric version of this simulation. The nongeometric version generates random sums, and provides a good way to understand the problem. The geometric version uses points to represent random sums; it simulates the experiment with a geometric model. After using the nongeometric version, you must return to the main menu in order to access the geometric version.

Key Commands The features here are identical to those described for the dartboard simulation. In the geometric model, pushing the space bar will pause the action and provide additional information, including the equation of the boundary line and the theoretical probability.

Interpreting the Screens The nongeometric version of this simulation displays randomly generated sums. The fact that these sums have been truncated may give the false impression that this is a discrete problem. Students may need to be reminded that x and y can assume any values between 0 and 5, and their sum can therefore take on any value between 0 and 10. The summary lists a variety of empirical probabilities. For different values of n, the screen displays the percent of sums less than n, as well as the percent between $n-1$ and n. These percents have been rounded to the nearest tenth. Note that since this is a continuous problem, the theoretical probability that the sum is less than n is identical to the theoretical probability that it is less than or equal to n.

In the geometric version, the user chooses a value for k, and the software finds an empirical probability that $x + y \leq k$. Since it is certain that $x + y \leq 10$, the value chosen for k should be less than 10. The screen shows the sample and event spaces for a geometric model of the random sums.

19 Random Quotients Simulation

Again, the user can choose between a geometric version and a nongeometric version. It is recommended that for classroom demonstrations the nongeometric version be used first.

Key Commands Refer to the explanation for these keys given previously. In the geometric version, pushing the space bar will pause the output and display some additional information, including the equation of the boundary line and the theoretical probability.

Interpreting the Screens The nongeometric version of this software generates random quotients and presents a summary of the distribution of these quotients. The quotients themselves have been truncated, and the percents given in the summary have been rounded to the nearest tenth.

In the geometric version of this simulation, the user chooses a value for k and the software uses a geometric model to find an empirical probability that $y/x \leq k$. The screen shows the sample and event spaces for the geometric model; the random quotients are represented by randomly chosen points.